Examining Canadian Energy

Examining

Natural Gas

Laura Clifford

SAUNDERS BOOK COMPANY

Published by Saunders Book Company
27 Stewart Road, Collingwood, ON Canada L9Y 4M7

www.saundersbook.ca

Produced by Red Line Editorial

Library and Archives Canada Cataloguing in Publication

Clifford, Laura, author
Examining natural gas / Laura Clifford.

(Examining Canadian energy)
Includes bibliographical references and index.
Issued in print and electronic formats.
ISBN 978-1-77092-272-3 (bound).--ISBN 978-1-77092-280-8 (pbk.).--
ISBN 978-1-77092-288-4 (html)

1. Natural gas--Canada--Juvenile literature. I. Title. II. Title: Natural gas. III. Series: Examining Canadian energy

TN882.C3C55 2014 j553.2'850971 C2014-903303-6
C2014-903304-4

Picture Credits

Shutterstock Images, cover, 1, 15, 18, 24, 28–29; Ulga/Shutterstock Images, 5; Jubal Harshaw/Shutterstock Images, 8; Dmitri Ometsinsky/Shutterstock Images, 9; Reinhard Tiburzy/Shutterstock Images, 11; Fotolia, 13; iStockphoto, 21, 37; Red Line Editorial, 23, 33; Doug Edgar/Thinkstock, 27; Thinkstock, 32; Keith Srakocic/AP Images, 34; Leonid Ikan/Shutterstock Images, 38; Željko Radojko/Fotolia, 41; Waverly Wyld/iStockphoto, 45

Printed in the United States of America
Corporate Graphics Inc.
July 2014

Contents

Chapter 1	**Natural Gas: For Yesterday, Today, and Tomorrow**	**4**
Chapter 2	**Prehistoric Power**	**7**
Chapter 3	**Natural Gas at Home**	**12**
Chapter 4	**The Chemistry of Natural Gas**	**17**
Chapter 5	**How We Find Natural Gas**	**20**
Chapter 6	**Getting, Processing, and Moving the Gas**	**26**
Chapter 7	**The Power of Water**	**31**
Chapter 8	**The Future of Natural Gas**	**36**
Chapter 9	**Your Turn**	**40**

Glossary **42**
Explore Further **44**
Selected Bibliography **46**
Further Information **47**
Index **48**

CHAPTER **1**

Natural Gas: For Yesterday, Today, and Tomorrow

Have you heard people talking about the high cost of energy? Does your family discuss the cost of heating your home? Right now, a lot of the world's energy comes from non-renewable sources. These non-renewable sources, such as oil, can have negative effects on the environment, and the sources will eventually run out.

Scientists are constantly looking for ways to improve our sources of energy. They want to use energy in ways that are more efficient, less expensive, and better for the environment than how our current energy sources are used. One source of energy we might make better use of is natural gas. People

Do you use natural gas at your house?

already use natural gas every day to heat their homes, cook, and even fuel their cars. Businesses and industries use it, too. Canada is the third-largest natural gas producer in the world, and we continue to find more ways to use this resource.

Natural gas is a fossil fuel. It forms from organic materials buried underground for millions of years. All fossil fuels cause pollution, and their supply is limited. But natural gas pollutes less than the other two fossil fuels, oil and coal. Right now, natural gas meets about 30 percent of Canada's energy demand. Some people believe we should increase the amount of natural gas that we use compared to our other energy sources.

EXPLORING NATURAL GAS

In this book, it is your job to learn about natural gas and its place in our energy future. Where does natural gas come from? When did people begin using it as fuel? What do we use it for today? What might we use natural gas for tomorrow?

Taylor Anderson is researching natural gas for a presentation for his science class. He is meeting with scientists and other energy innovators who will help him learn more about natural gas and how it can help meet our energy needs. Reading Taylor's journal will help you conduct your own research.

CHAPTER 2

Prehistoric Power

I learned from Ms. Schuler in science class that it's good to come up with questions when researching something. I wrote down two questions for my research today: Where does natural gas come from? When did people start using natural gas as a fuel?

I decide to start by interviewing Ms. Schuler herself after school one day.

"Natural gas has been around for a long time," she tells me. "Most of the natural gas we find today is millions of years old, so it is buried deep inside the earth."

She explains that natural gas is a fossil fuel, like oil and coal. Fossil fuels are made from organic material.

"Believe it or not," she says, "natural gas actually comes from sea creatures. Tiny plants and animals lived hundreds of millions of years ago. When the plants and animals died,

Natural gas comes from ancient creatures that have been buried underground for hundreds of millions of years.

they ended up at the bottom of the ocean. Over a really long time—many millions of years—layer upon layer of sand and silt covered the organic remains. This buried them deeper and deeper in the earth. There, heat and pressure changed the buried organic matter into oil, coal, and gas."

Ms. Schuler tells me that natural gas has not only been around for a long time but also that people have been using it for thousands of years. People living in the Middle East discovered natural gas seeps in Iran between 6000 and 2000 BCE. Seeps are spots where the fuel leaks through the earth and escapes. Sometimes, lightning would set the gas on fire. Because the gas was seeping from the earth, it would continue burning, sometimes for years. People who saw the fire didn't understand what was happening. In some cultures, people thought these fires were magical or divine. They became an important part of religions in Persia (present-day Iran), India, and Greece. A famous temple in Delphi, Greece, was even built around one of these flames.

Ms. Schuler says, "In 211 BCE, the Chinese created the first known natural gas well. They drilled 150 metres into the earth and then inserted bamboo poles into the well. Because bamboo is naturally hollow, the poles directed the fuel where the Chinese wanted it to go."

I learn that people in the Western Hemisphere discovered and used natural gas much later than in China or the Middle East. In 1659, natural gas was discovered in England. But the English did not have much use for natural gas.

"At least not right away," Ms. Schuler quickly adds. "They began using

Ancient Greeks considered the natural gas seeps in Delphi to be sacred.

CONVENTIONAL AND UNCONVENTIONAL

Natural gas exists across Canada, though the largest reserves are found in the western part of the country. Canada's natural gas comes from both conventional and unconventional sources. Conventional sources are easier to access and account for most of the country's natural gas production to date. They include porous rocks, such as sandstone and carbonates, that allow gas to move more freely to wells. About 91 percent of Canada's natural gas comes from conventional sources. Unconventional sources are found in less porous rock formations, making them more difficult to produce. These types of reservoirs include shale gas, coal bed methane, and gas hydrates. New technologies are being used to access these unconventional sources and make them available for use.

gas in homes and in streetlights in 1785. But that was a gas manufactured from burning coal, not the gas that comes from the ground."

Ms. Schuler goes on, "One of Canada's first onshore natural gas discoveries occurred in New Brunswick in 1859. But it was not sold for commercial use. Instead, it was flared, or burned as a waste product. About seven years later, natural gas was discovered in southern Ontario. By 1889, it was being used for heating and lighting. In 1883, a Canadian Pacific Railway crew was digging for water near Medicine Hat, Alberta, when they discovered natural gas. By the 1890s, natural gas from this area was being used to supply homes and factories."

But, Ms. Schuler tells me, there was still the challenge of getting natural gas to people who didn't live near natural gas

Natural gas pipelines helped make natural gas a more practical energy source.

wells. The first Canadian pipeline was built in 1853 to carry natural gas to Trois Rivières, Quebec. At 25 kilometres, it was possibly the longest pipeline in the world at the time. Over several decades, Canadian companies expanded the number of pipelines connecting communities across the country. One of the most significant early pipelines was built in Alberta in 1912. It was 275 kilometres long, and it connected Bow Island to Calgary. The pipeline made gas more accessible to more people.

Then, in the late 1960s and early 1970s, an oil shortage led to a worldwide energy crisis. People began looking for other fuels. That's when natural gas really took off as an energy source. Today, more than 40 percent of Canadian homes use natural gas as their main energy source. In addition, nearly half of Canadian homes, mainly in Ontario and Western Canada, use natural gas as their main heating fuel.

I think I've got a really good start on my project. Next, I'll look at how we use natural gas.

CHAPTER 3

Natural Gas at Home

I learned today just how much people rely on natural gas as an energy source. People all over the world use it every day. My mom is an engineer for the local gas company. When I get home from school, I decide to ask her how people today use gas in their own homes.

I know our stove uses natural gas. But I'm surprised when she tells me the clothes dryer uses natural gas, too.

"I didn't know that," I say. "Is there anything else?"

"Well, the house is nice and toasty in the winter thanks to natural gas. And you can thank natural gas for powering the water heater so you can take hot showers."

I think I know another one. "Isn't the grill on the deck fueled by natural gas?"

"It is," Mom confirms.

"Are there any other things in the house that use natural gas?" I ask.

"Not in our house," she replies, "but your aunt Sarah has a gas fireplace."

Of course! I remember that she doesn't have to use firewood or matches. She just flips a switch to start a fire.

Natural gas fuels much of the technology we rely on every day.

My mom tells me natural gas makes life comfortable in more places than just homes. Many schools and businesses also use natural gas to heat their buildings and their water. Natural gas can also be used to cool buildings with air conditioning. Natural gas powers things like lights and computers as well. Some companies use natural gas to provide energy to power factories. Almost 9 percent of natural gas used in Canada is used for generating electricity.

"We rely on natural gas for so much more than energy," Mom says. "Natural gas goes into making products such as antifreeze, fabric, plastic, and fertilizer."

"Fertilizer?" I ask. "Natural gas helps things grow?"

"It does!"

Mom explains there's a procedure chemists use to create fertilizer from methane, the main ingredient in natural gas. This process changes natural gas into ammonia, which is developed into fertilizer. Scientists estimate at least one-third of all people in the world rely on food produced using these fertilizers.

"And our uses for natural gas are still growing," my mom tells me. "Natural gas is starting to be used for transportation, and the government is helping make that happen."

She explains that most vehicles run on fossil fuels—the gasoline created from oil. Gasoline causes pollution when burned in a car or bus. The pollution releases carbon dioxide into the atmosphere. Some scientists believe that too much carbon dioxide traps the sun's energy like heat in a greenhouse, which could lead to a gradual increase in the earth's temperature. Natural gas releases carbon dioxide when it is burned, but it gives off much less carbon dioxide than other fossil fuels, making it a cleaner source of energy. Natural gas vehicles (NGVs) are better for the environment.

"How is the government encouraging NGVs?" I ask.

This three-wheeled taxi in Thailand runs on compressed natural gas.

"They're providing grants," Mom says. "The government is giving groups money to invest in projects that use alternative fuels. In Hamilton, Ontario, many city buses are fueled by natural gas.

"And that's not the only benefit to natural gas–powered cars," my mom adds. She tells me NGVs save operators 40 to 60 percent in fuel costs compared to traditional gas-powered

NATURAL GAS VEHICLES

Natural gas helps fuel transportation. Natural gas is less expensive than gasoline and diesel, and it emits fewer pollutants than these fuels when burned. Almost 15 million vehicles around the world run on natural gas. Natural gas vehicles (NGVs) use either compressed natural gas (CNG), which is natural gas under pressure, or liquefied natural gas (LNG), which is compressed gas that has been cooled to make it become a liquid. LNG vehicles are good for driving long distances. CNG vehicles are well suited to shorter distances. They serve as taxis, postal trucks, street sweepers, and more. In 1985, the Hamilton Street Railway (HSR) worked with various partners, including the Province of Ontario and Natural Resources Canada, to develop the first compressed natural gas (CNG) transit buses the world had ever seen. The majority of funding for the project came from government grants, which were used to convert seven diesel buses into NGVs. Soon after HSR started using the CNG buses, other Ontario cities, such as Toronto and London, did the same. Today, there are more than 250,000 CNG buses around the world that were developed using the HSR technology, which is considered the world's leading alternative bus technology.

vehicles. There are many natural gas fields in Canada, and new technology has given us new ways to mine them. NGVs could help create jobs in this new energy field and reduce our overall carbon emissions.

I think I understand how we use natural gas today. But I'm still not sure exactly what natural gas is or how we turn it into energy.

CHAPTER 4

The Chemistry of Natural Gas

Today, I am meeting with Dr. Elizabeth Roberts, a chemist at a university in British Columbia. Dr. Roberts studies what things are made of and how they interact. I explain to her that I know where natural gas comes from, but I want to know exactly what it is made of.

"Natural gas is a combination of things," says Dr. Roberts. "It's mostly methane, which is a type of gas called a hydrocarbon. Hydrocarbons are made of only two elements: hydrogen and carbon."

She points to a chart on the wall that's called the periodic table of the elements. The table has many squares. Each square has one or two letters, plus some numbers. The table has different colours, too. Hydrogen and carbon are two of the squares on the table.

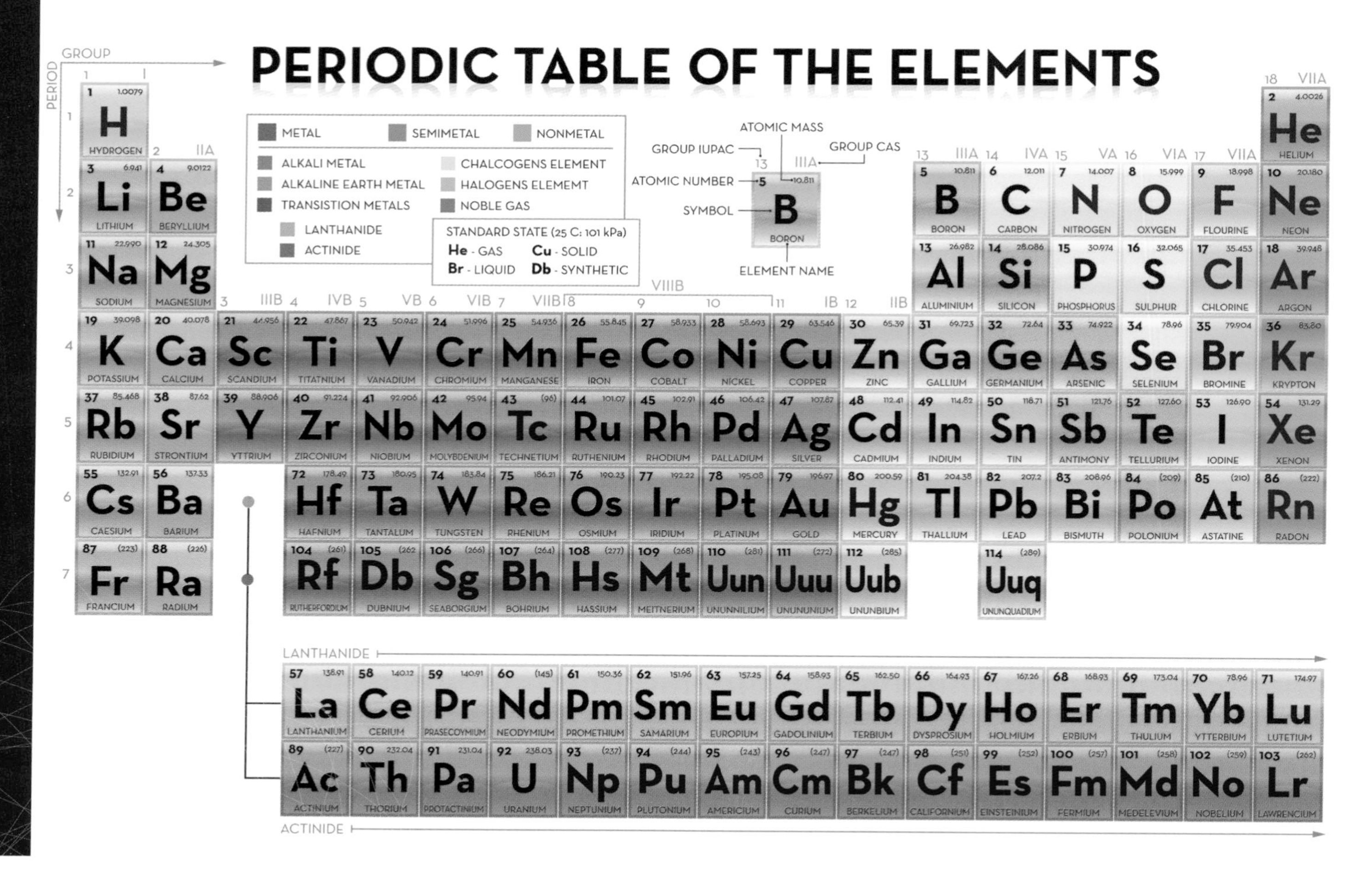

The periodic table of the elements lists information about every known element. An element is the most basic form of a substance.

"Are there other kinds of hydrocarbons in natural gas?" I ask.

"Yes," Dr. Roberts says. "Ethane, butane, and propane are often found in natural gas."

"So, natural gas is a mixture of hydrocarbons," I confirm.

"Yes," she replies. "But it has non-hydrocarbons, too. Other gases often exist with the hydrocarbons, including carbon dioxide, helium, hydrogen, and nitrogen."

"Wow," I say. "Natural gas really isn't one thing. It's lots of things."

Next, Dr. Roberts talks about some of the physical properties of natural gas. I already know that natural gas is flammable—it catches fire when my mom turns on a burner on our stove.

Dr. Roberts also tells me natural gas is odourless and colourless. You can't see it or smell it. It also doesn't create any smoke when it burns. Dr. Roberts says there's a lot more to know about natural gas. She suggests I meet with a geologist, someone who studies rocks and the earth.

H_2O

Chemists use formulas to describe substances. For example, you may have heard of "H-2-O." That's water, and H_2O is its chemical formula. The formula tells us there are two hydrogen (H) atoms and one oxygen (O) atom in a molecule of water. Natural gas is mostly methane and ethane. Since they are hydrocarbons, you know they are made of only hydrogen and carbon (C). The chemical formula for methane is CH_4. The formula for ethane is C_2H_6. What does that tell you?

CHAPTER 5

How We Find Natural Gas

Now I know where natural gas comes from, a bit about what it's made of, and some of its properties. Next, I want to learn how we find natural gas.

Today, I am meeting with Dr. Chris Phillips, a geologist in Manitoba. His office has a wall full of shelves covered with books and rocks. He also has math books and all kinds of science books on geology, chemistry, and physics.

"Wow," I say. "You sure know a lot about science."

"Yes," he says. "I had to study a lot of things to become a geologist."

Dr. Phillips starts by telling me a little bit about natural gas. "Natural gas is classified one of two ways," he says. "It is either associated or non-associated. Associated gas is found with oil. The oil and gas are associated—they are companions.

This type of natural gas often contains liquids, so it's also called wet gas."

Dr. Phillips then explains that non-associated gas isn't associated with oil. Because it occurs by itself and doesn't contain liquids, non-associated gas is also known as dry gas. He draws an example of non-associated gas. Sometimes it's within sand or shale, a type of rock. Just like associated gas, non-associated gas is often deep in the earth.

Dr. Phillips tells me that sometimes there's evidence of natural gas below ground. Not long ago, scientists found natural gas by looking for seeps, just like those people saw hundreds and thousands of years ago. But Dr. Phillips says finding fuel deposits this way isn't very common anymore. Today, technology helps locate natural gas. When he

Some geologists look for rock formations likely to contain natural gas.

goes out into the field to look for evidence of natural gas, he is not looking for seeps.

"What do you look for?" I ask.

"I look for a land formation called an anticlinal slope," Dr. Phillips says. "It's a place where the earth folded."

"You mean like how a shirt is folded?" I ask.

"Exactly," said Dr. Phillips. "When this happens, a dome results." He pulls a book from one of the shelves and flips through the pages. He shows me a page with a photograph and an illustration of an anticlinal slope. The earth really had folded. "History has shown that these slopes are more likely than other places to have reservoirs of oil and natural gas," he adds.

Dr. Phillips tells me that to find natural gas, scientists also rely on seismology, which is the study of earthquakes and man-made vibrations in the earth. Dr. Phillips shows me seismograms, the records of some of these vibrations. One was basic, with squiggly lines. The other was more complicated, with layers in different colours. He says there are even three-dimensional seismograms.

Dr. Phillips says small instruments called geophones are often set up in an area being studied for potential natural gas. Geophones measure vibrations created by humans. Scientists used to explode dynamite in the ground to cause the vibrations. Today, they use big trucks. The heavy trucks create vibrations when driven over the area, and the geophones measure the vibrations to see if natural gas is present.

"The waves bounce off different rocks in different ways," says Dr. Phillips. "Think of bouncing a ball. It will bounce one way on a sidewalk and another way on grass."

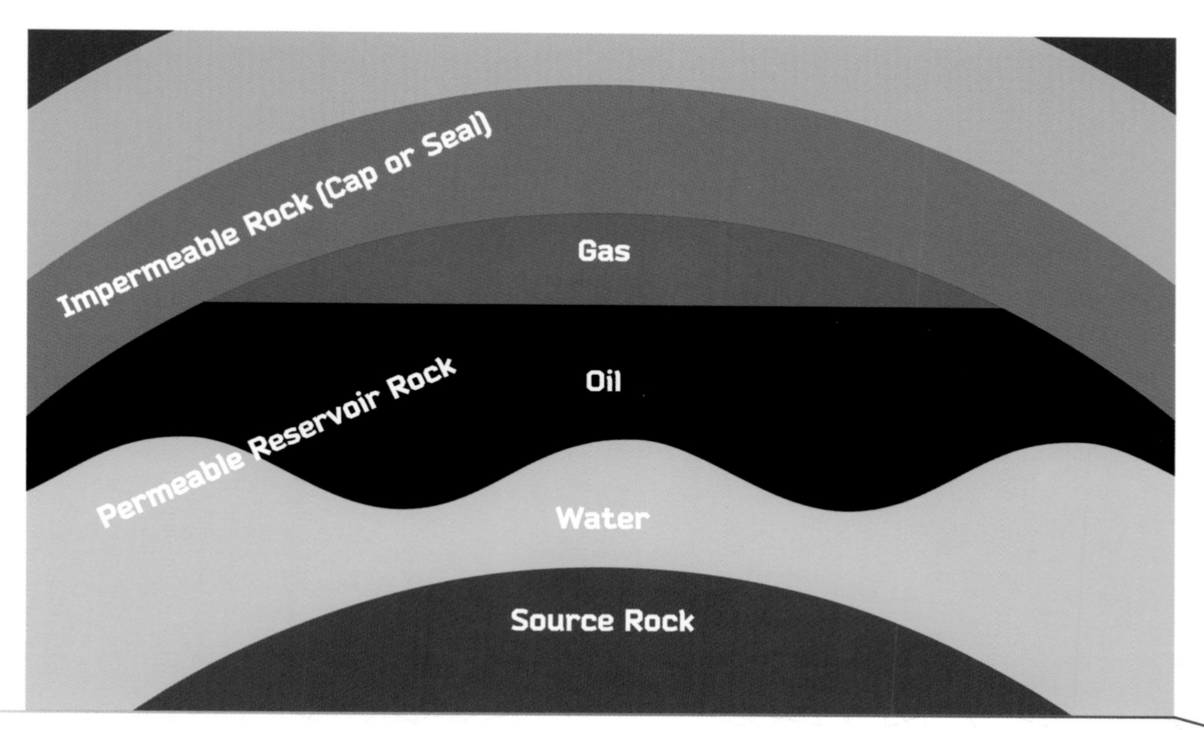

NATURAL GAS ROCK LAYERS

Natural gas is often found deep in the earth in a rock formation called a gas trap. The trap's arched shape makes it possible for natural gas to form. A gas trap has several layers of rock. The source rock is the source of the natural gas. The reservoir rock is the rock that natural gas seeps into. Although the reservoir rock is solid, it is also porous so gas can flow through it. A layer of water, oil, and gas sits between the source rock and the reservoir rock. The cap rock is at the top of the formation. The cap rock is less porous than the reservoir rock, so it holds everything into the rock. It's also called a seal.

Shale rock often contains natural gas.

"So, you don't measure natural gas, you measure vibrations against rocks?" I ask.

"Right," he confirms. "The seismogram helps identify where traps might be."

I can understand how these geophones work on land, but I have heard that a lot of natural gas is mined from under the ocean. I wonder how scientists find that kind of gas.

"Can you use seismology to find natural gas deposits in the ocean?" I ask.

"That's a great question," says Dr. Phillips. "We use the same process with equipment suitable for water. A boat pulls an instrument that shoots compressed air into the water. Another boat follows, pulling geophones that work in water, known as hydrophones. The hydrophones pick up seismic waves, letting us know if there is natural gas beneath the ocean floor."

Dr. Phillips also tells me he uses instruments that measure the earth's magnetic and gravitational fields when he is seeking natural gas. This information helps Dr. Phillips identify what types of rocks are underground. A computer processes the information from these sources and creates a picture of the geology of the earth below. Dr. Phillips interprets the picture. He also says a fuel company that thinks it has found a natural gas deposit will usually drill an exploratory well. That's really the only way to know for sure if natural gas exists in a location. If natural gas is discovered, the well will be developed for production.

More goes into finding natural gas than I thought. But I wonder what we do with the gas once we find it.

THE DEVELOPMENT OF SEISMOLOGY

Seismology was first used to search for natural gas and oil in 1921. Before that time, it was used to measure earthquakes, something scientists still use it for today. Chang Heng, a Chinese scholar, developed the seismoscope in 132 CE. It could detect movement kilometres away from the seismoscope's location. In 1855, Italian scientist Luigi Palmieri invented a seismometer that measured ground motion, as well as the time and intensity of the movement. Seismology continued developing into the technology geologists and other scientists use today.

CHAPTER 6

Getting, Processing, and Moving the Gas

Now that I understand how scientists detect natural gas, I want to learn more about how they get it out of the ground. Today, I'm in Alberta to meet with Rajan Palami, an engineer with an energy company. He oversees drilling at the well I'm visiting.

When I arrive, Rajan hands me a helmet and safety glasses. "Welcome," he says. "I hear you want to learn about natural gas production."

Rajan explains that vertical wells are made using rotary drilling. A giant drill bit is spun into the ground. "Have you ever drilled a hole in a piece of wood?" Rajan asks.

"We're doing the same thing but on a much larger scale."

He explains that a device called a mud motor is located just above the drill bit, or tip of the drill. The mud motor turns the bit.

"What keeps the sides of the hole from caving in?" I ask.

"There are a few steps to finishing a well," Rajan says. "First, metal well casing is inserted into the well. The casing has several layers, and it strengthens the sides of the well. It also keeps gas from seeping out and other matter from getting in. Next, a wellhead is added. This is the part of the well you can see aboveground. It's also made of metal. The wellhead controls extraction and keeps leaks from happening."

The wellhead keeps the gas from escaping from the well and allows miners to access the well when they are ready.

"So, now the well is ready to pump out the natural gas?" I ask.

Rajan says the well we're at pumps out dry gas. This gas flows on its own because it's lighter than air—it simply needed to be accessed by the well to escape. Wet gas comes from an oil well, which uses special extraction equipment.

Rajan says some wells are located in water, often in the ocean. This is called offshore drilling. Drilling in the ocean is similar to drilling on land. The energy company usually builds a platform for the drilling rig to sit on. The platform is connected to an underwater well. Some platforms are fixed and can't be moved, while others can move from place to place.

"Not all offshore drilling occurs in oceans," says Rajan. "Drilling barges are large platforms that float. They are made

Natural gas is cooled until it becomes a liquid that ships can transport overseas.

for shallow waters, such as lakes and rivers. Tugboats pull them from one location to another."

Next, we talk about producing natural gas. All natural gas is produced the same way. Where it was found or how it was obtained doesn't make a difference.

"Once natural gas is retrieved," says Rajan, "it needs to be refined, or cleaned up. The natural gas is refined to make it mostly methane. Impurities, such as water and nitrogen are removed, as well as some hydrocarbons. Some of these hydrocarbons can be used independently of the natural gas. Propane is an example. It can be used as fuel for engines, barbecue grills, and central heating, and to create plastic and synthetic fabric."

DO YOU SMELL ROTTEN EGGS?

By providing energy, natural gas is useful to people and businesses, but it can also be harmful. Natural gas is explosive. Breathing in too much can also poison animals and humans. Because natural gas has no colour, odour, or taste, people can be unaware when there is a gas leak. A spark could easily turn a leak into an explosion. That's why utility companies add the chemical called mercaptan to natural gas during the production process. It makes natural gas smell bad—like rotten eggs.

Once it is refined, natural gas is ready for use. In Canada, Rajan explains, transporting natural gas involves a 500,000-kilometre network of pipelines.

"These pipelines take natural gas from the well to the processing plant and then to where the natural gas is needed—usually cities. Ships take natural gas to destinations across the globe. Before shipping, the gas is cooled until it becomes a liquid. Natural gas takes up a lot less space as a liquid, which makes it easier to transport."

Rajan explains that the natural gas goes into storage or to the utility company. The utility company has its own system of pipelines that distribute natural gas to customers.

Rajan says a new type of drilling called hydraulic fracturing, or fracking, is providing energy companies greater opportunities to access natural gas. I've learned so much from my visit and am eager to find out more.

CHAPTER 7

The Power of Water

Today, I'm visiting a hydraulic fracturing operation outside of a small town in Quebec. Environmental engineer Trevor Burns has offered to show me around the site. He tells me that hydraulic fracturing is being used more and more as a means of getting natural gas.

"Not too long ago," Trevor says, "scientists were worried that Canada was going to run out of natural gas. Although hydraulic fracturing has been around for more than 60 years, the technology has improved. Today, hydraulic fracturing allows us to access natural gas we would otherwise have no way of reaching. The bedrock under land near Montreal and Quebec City and a few other places near the Appalachian mountainfront is part of a giant shale formation. Areas of British Columbia, Southern Alberta, Saskatchewan, New Brunswick, Ontario, Manitoba, and Nova Scotia also have shale formations. This

Hydraulic fracturing wells are used to access natural gas in British Columbia.

shale is full of trapped natural gas. Using hydraulic fracturing, we are able to mine the natural gas."

"Is hydraulic fracturing different from the other kinds of drilling?" I ask.

"It is," he says, "though part of it looks the same." Trevor points to a tall tower not too far in the distance. "Hydraulic fracturing uses a well, just like other forms of drilling. But this well is different. Instead of reaching straight down into the ground, hydraulic fracturing wells run down then bend to the side. The well can run horizontally for several kilometres."

He tells me tools near the drill bit take, or log, measurements and send the information to the surface. This is called logging while drilling (LWD), and it provides information about the rock being drilled, such as its porosity. Trevor says a type of LWD called measurement while drilling (MWD) sends information about direction, temperature, pressure, and the drill bit. Workers use MWD and LWD to adjust drilling. The information helps them navigate underground.

Trevor explains that once the well is dug, millions of litres of liquid are pumped into it. The liquid is approximately 95 percent water, 4.5 percent sand, and 0.5 percent chemicals.

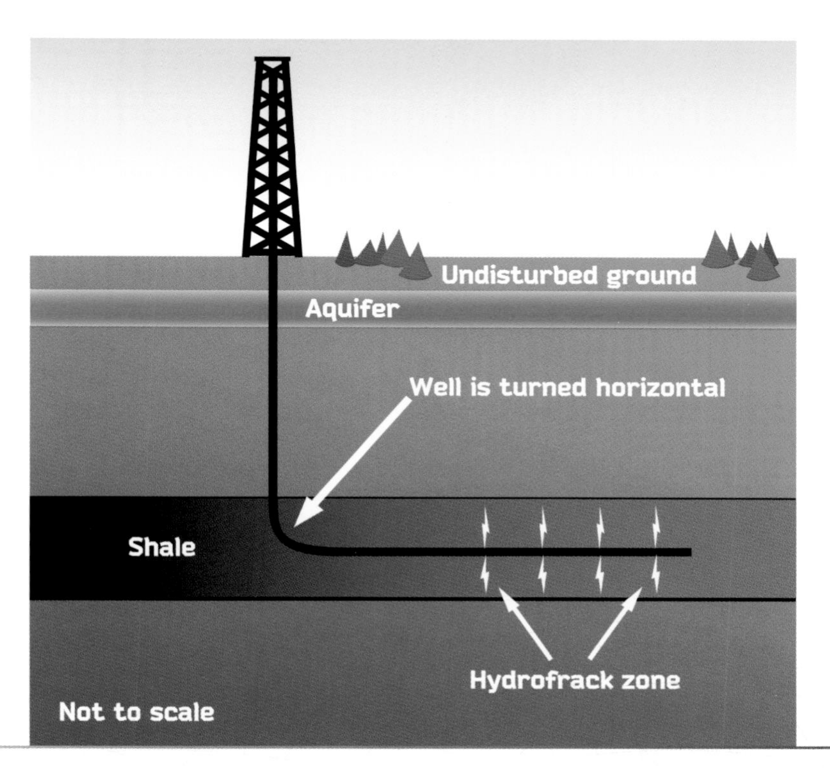

DIRECTIONAL DRILLING

Traditional gas wells are vertical, but not all drills are straight lines. Directional drilling takes wells in new directions, allowing them to reach locations not directly below. This is extremely helpful when natural gas exists below a railroad or protected nature area, where building a rig isn't possible. Directional drilling isn't new, but technology has made it better. Older wells took about 600 metres to bend from vertical to horizontal. Today, a well changes direction—even 90 degrees—in only a few metres. Advances in drilling technology have made more natural gas deposits accessible. They also make drilling more economical, especially offshore, where one rig can have 20 or more directional wells.

A worker creates the liquid mixture used in hydraulic fracturing.

All that liquid is forced into the well. The watery mixture pounds the shale and causes fractures, or cracks, in it. The natural gas escapes through those cracks and can be collected by the well.

"This method is extremely effective and relatively inexpensive. It has led to a lot of new jobs in Canada," Trevor says. "Canada has huge deposits of shale. If we are able to access the natural gas in this shale, it could go a long way toward achieving energy security. Hydraulic fracturing also works out well for the surface environment because it doesn't disturb the surface as much as drilling in vertical wells does. But hydraulic fracturing has its problem, too," Trevor adds.

Trevor tells me that a lot of people are worried about the amount of water being used in hydraulic fracturing. Using up so much water can change the landscape and affect the wildlife living in that environment. Plus, people in the towns nearby need the water for drinking. Many people are concerned that the chemicals used in hydraulic fracturing can get into drinking water because they seep into the ground and could get into the water supply. Also, the contaminated water used in hydraulic fracturing must be disposed of in a safe manner or treated to make it reusable.

Because hydraulic fracturing is a relatively new technique, Trevor says, we are studying the safest methods to extract gas from the ground without harming the environment. "Fracking has potential," he says. "But unlike other means of getting energy, it has a ways to go before we can understand its risks."

DIGESTERS: ANOTHER KIND OF NATURAL GAS

Not all natural gas comes from underground. Some farms and sewage treatment plants, where waste is processed from homes and other buildings, use digesters to create energy. Digesters are systems in which tiny living creatures digest animal or food waste. Methane is produced, along with carbon dioxide and hydrogen sulfide. Instead of taking millions of years, this process only takes a few weeks. The methane can be burned for electricity or processed for commercial natural gas lines. Digesters release pollutants, and they can be smelly because the material being processed is sewage or rotting food.

CHAPTER 8

The Future of Natural Gas

For my last stop, I am back in British Columbia, meeting with Rebecca Wallis. She is a professor of geochemistry at the local university and is interested in the future of natural gas.

Thanks to my research, I understand there are a lot of pros and cons to natural gas as an energy source. One of the biggest cons is that natural gas is a non-renewable fossil fuel—we are going to run out of it at some point. I ask Professor Wallis how long our natural gas will last.

"Well, that depends on who you ask," Professor Wallis responds. "Some experts think natural gas could be the country's new frontier in energy. Canada produces more natural gas than it needs each year, so it exports leftovers to the United States."

Pipelines transport natural gas around the world. Because of improvements in mining techniques, we are using more natural gas than ever before.

Professor Wallis tells me scientists are unsure just how long our supplies will last. Canada appears to have an abundant supply of natural gas. But experts say our natural gas stores will last about 100 years because technology has improved the ways in which we locate and drill for natural gas. Hydraulic fracturing is a big part of that estimate.

Professor Wallis says another source of natural gas is being studied. Gas hydrates are solids made of gas and water molecules. They are located in out-of-the-way places such as in the Arctic and below the ocean floor. An example is the Mallik gas hydrate field found at the northeastern edge of Canada's

Some oil mining operations burn off natural gas rather than mining it.

Mackenzie Delta. It is one of the most concentrated gas hydrate reservoirs researchers have found to date.

"This is just one example, and gas hydrates are still being studied," Professor Wallis explains. "Scientists aren't sure how much of the methane in gas hydrates is actually recoverable. Physically getting it could be very challenging. Time will tell, of course."

Professor Wallis continues, "In a lot of ways, natural gas is a better energy source than coal and oil. Natural gas contributes the least to pollution. It is the cleanest of the fossil fuels. And there is a lot of it available right in Canada. Today we are using more natural gas than ever before. Natural gas adds a combined total of about $450 billion to the U.S. and Canadian economies each year. The natural gas industry employs about 3.4 million people between these two nations, and increasing natural gas production will likely help the economy grow."

She pauses for a moment before saying, "On the other hand, natural gas still pollutes the environment and is non-renewable. Also, the methods used to get natural gas might harm the environment and people."

Professor Wallis has given me a lot to think about. Natural gas is an important fuel. The natural gas industry provides jobs and money for people, businesses, and countries. But it's important to find a way to get the fuel without harming the environment. Natural gas is an amazing resource. We need to continue finding ways to use it wisely.

FLARING

Not all natural gas that is drilled is processed. Instead of capturing the gas that occurs with oil, some energy companies around the world burn it off in a process called flaring. The gas is worth less than the oil, so oil companies flare gas instead of capturing it. If companies could store and transport the gas economically, many homes and businesses could be fueled by it. Flaring is a major concern in Alberta, where oil and gas is one of the province's main industries. Residents of communities such as Cochrane and Lochend are concerned about health issues arising from nearby flaring. In the late 1990s and early 2000s, Alberta oil and gas companies partnered with the Clean Air Strategic Alliance (CASA) to reduce emissions from flaring by up to 70 percent. However, since then, rates once again have increased, and community groups have formed to speak out against the issue.

CHAPTER 9

Your Turn

You've had a chance to follow Taylor as he conducted his research. Now it's time to think about what you've learned. Natural gas is developed from the remains of creatures that lived and died millions of years ago. Natural gas consists mostly of the hydrocarbon methane, which is colourless and odourless. Natural gas is a good energy option because it pollutes less than other fossil fuels and is available in Canada, providing jobs and helping the country move toward energy security. However, some techniques used to retrieve natural gas are controversial and might harm the environment. Natural gas is also a non-renewable resource; someday it will run out. While alternative energy research has made progress, it still has a long way to go. Maybe you will be part of the solution!

YOU DECIDE

1. Natural gas has been a reliable resource for a long time. Should people continue to rely on it or explore other energy options?
2. What do you think about energy companies exploring new deposits of natural gas?
3. What can you do to cut down on your energy use and preserve our energy resources? Think about technology and also ways to change your behaviour, like walking instead of riding in the car.
4. If you were responsible for choosing the energy sources your family uses and paying for them, which would you choose? Why?
5. The decision to increase production of natural gas has pros and cons. What do you think is most important to consider when it comes to producing natural gas? Explain your reasoning.

How big a role do you think natural gas should play in our energy future?

GLOSSARY

alternative energy: A type of energy that is from a renewable source that is not in danger of running out, such as the sun, wind, or water.

atom: The smallest particle of an element.

carbon: An element in fossil fuels such as coal, oil, and natural gas.

carbon dioxide: A byproduct that is created by the burning of fossil fuels.

flammable: An object that is able to catch fire and burn.

fossil fuel: An energy source that developed from the remains of animals and plants.

hydraulic fracturing (fracking): A relatively new technique to extract oil and natural gas from the ground, especially from shale formations.

hydrocarbon: An organic compound made of carbon and hydrogen.

methane: A gas with no colour or odour that is produced by decaying organic matter.

non-renewable: Something that cannot be replaced by natural environmental cycles as quickly as we are using it.

organic: Material from living plants or animals.

porous: Having pores or holes.

renewable: Something that can be replaced by cycles in nature or the environment.

reservoir: A place where something is collected, especially a liquid.

seismogram: A graph showing shaking in the earth.

shale: A type of rock formed from clay, mud, or silt that often contains natural gas.

silt: Tiny particles of dirt carried by water.

EXPLORE FURTHER

Map Your Natural Gas Usage

If your family uses natural gas at home, such as for heating or cooking, ask your parents if you may see some recent gas bills. Make a simple chart that shows the money spent on your gas bill each month. The bills may already include such a chart. Examine the ups and downs of the chart. That's your usage. Discuss with your parents why your family's use of natural gas changes from month to month or season to season. Next, talk about what you can do individually and as a family to save money on your gas bill. Working together, you can save energy and money.

Explore Porosity

You learned that natural gas seeps into reservoir rock because the reservoir rock is porous. You also learned the cap rock traps gas because it is denser and lacks porosity. Explore porosity with this simple experiment. You'll need sand, clay, two jars, and water. First, compare the sand and clay. Look at them. Touch them. Next, fill one jar almost to the top with sand. Fill the other jar almost to the top—so it's level with the first—with clay. Then, fill both jars to the top with water. Did you pour more water into one jar than the other jar? What does that tell you about the sand and the clay?

Recycle!

One way to extend the energy supply is to limit the need for it. In addition to using less natural gas to heat and cook, you can recycle. Recycling is important in preserving natural resources. Using products again reduces the need to make new products, including those in which natural gas is a key ingredient, such as plastic. Learn about recycling programs in your area. What can you recycle? Does the city have a recycling program that has a regular pick-up schedule, or are there recycling centres where you can bring items? If your family doesn't recycle, make a plan to start recycling! If you do recycle, find out if there's more you can do.

Recycling can be a great way to conserve resources.

SELECTED BIBLIOGRAPHY

Bott, Robert D. *Evolution of Canada's Oil and Gas Industry*. Canadian Centre for Energy Information, 2004. Web. Accessed March 16, 2014.

Canadian Association of Petroleum Producers. *The Facts On: Natural Gas*. CAPP, 2014. Web. Accessed March 15, 2014.

Canadian Natural Gas. *Canadian Natural Gas: Introduction*. Canadian Association of Petroleum Producers, 2014. Web. Accessed March 15, 2014.

Milner, Alicia. "Hamilton Street Railway – The World's First CNG Transit Fleet." *Canadian Natural Gas Vehicle Alliance*. CNGVA, 2014. Web. Accessed March 14, 2014.

"Shale Gas." *Natural Resources Canada*. Government of Canada, 2014. Web. Accessed March 14, 2014.

FURTHER INFORMATION

Books

Benduhn, Tea. *Oil, Gas, and Coal (Energy for Today).* Pleasantville, NY: Gareth Stevens, 2009.

Flounders, Anne. *Power for the Planet (Our Green Earth).* Minneapolis, MN: Lerner Publishing Group, 2014.

Horn, Geoffrey M. *Coal, Oil, and Natural Gas.* New York: Chelsea Clubhouse, 2010.

Masters, Nancy R. *How Did That Get to My House? Natural Gas.* Ann Arbor, MI: Cherry Lake, 2009.

Websites

http://www.cga.ca
The Canadian Gas Association website has news and information about Canada's natural gas industry.

http://www.canadiannaturalgas.ca
The Canadian Natural Gas website explores everything you ever wanted to know about Canada's natural gas use, locations, and supply.

http://www.ngridenergyworld.com/ngsw/html/kids1.html
This website about natural gas includes experiments geared toward students.

INDEX

anticlinal slopes, 22
associated gas, 20–21

carbon, 16, 17, 19
carbon dioxide, 14, 18, 35
compressed natural gas (CNG), 16

directional drilling, 33
drilling, 9, 25, 26–28, 30, 32, 33, 34, 37, 39
dry gas, 21, 28

ethane, 18, 19

fertilizer, 14
flaring, 10, 39
fossil fuels, 5–6, 7, 14, 36, 38, 40

gas fields, 16
gas hydrates, 10, 37–38
gas traps, 23
geologists, 19, 20, 25
geophones, 22–25

hydraulic fracturing, 30, 31–35, 37
hydrocarbons, 17–18, 19, 29
hydrogen, 17–18, 19, 35

liquefied natural gas (LNG), 16

methane, 10, 14, 17, 19, 29, 35, 38, 40

natural gas, uses of, 4–5, 6, 7, 9, 10, 11, 12–16, 29–30
natural gas vehicles (NGVs), 14–16
nitrogen, 18, 29
non-associated gas, 20–21
non-renewable energy sources, 4, 36, 39, 40

offshore drilling, 28–29, 33

pipelines, 10–11, 30
pollution, 5–6, 14, 38
propane, 18, 29

refinement, 29–30

seeps, 8–9, 21–22
seismology, 22, 24, 25
shale, 10, 21, 31, 32, 34

wells, 9–11, 25, 26–28, 30, 32–34
wet gas, 21, 28